I love chocolate!

How is it made?

Cocoa beans grow.

A farmer plants many cocoa trees.
Hard **pods** grow on each tree.
Inside each pod are seeds called
cocoa beans.

FROM **COCOA BEAN** TO **CHOCOLATE**

by Robin Nelson

Lerner Publications Company / Minneapolis

Lerner Publications Company
A division of Lerner Publishing Group, Inc.
241 First Avenue North
Minneapolis, MN 55401 U.S.A.

Website address: www.lernerbooks.com

Library of Congress Cataloging-in-Publication Data

Nelson, Robin, 1971–
    From cocoa bean to chocolate / by Robin Nelson.
       p.   cm. — (Start to finish)
    Summary: An introduction to the process of making chocolate, from the time the farmer plants a cocoa tree to the time someone eats a piece of candy.
    Includes index.
    ISBN-13: 978–0–8225–4665–8 (lib. bdg. : alk. paper)
    ISBN-10: 0–8225–4665–5 (lib. bdg. : alk. paper)
    1. Confectionery—Juvenile literature.
    2. Chocolate—Juvenile literature. [1. Chocolate.]
    I. Title. II. Start to finish (Minneapolis, Minn.)
    TX792 .N45 2003
    641.3'374—dc21   2002000446

Manufactured in the United States of America
4 5 6 7 8 9 – DP – 13 12 11 10 09 08

The photographs in this book appear courtesy of: © Todd Strand/Independent Picture Service, cover, pp. 3, 23; © Wolfgang Kaehler, pp. 1 (top), 5, 9, 17; © Richard Nowitz, pp. 1 (bottom), 15; © Panos Pictures, p. 7; © Richard Nowitz/CORBIS, p. 11; © Corbis Royalty Free Images, p. 13; © Owen Franken/CORBIS, p. 19; © Annebicque Bernard/CORBIS Sygma, p. 21

Table of Contents

5

Workers open the pods.

The pods grow for many months.
Workers cut the pods from the
trees. The workers open the
pods with a large knife. There
are 20 to 50 cocoa beans inside
each pod.

The sun dries the beans.

The cocoa beans are taken out of the pods. Then they are left in the sun to dry for many days. The dry beans are put into large sacks.

A train takes the beans to a **factory**.

A train takes the sacks of cocoa beans to a chocolate factory.
A factory is a place where things are made.

The beans are roasted.

The beans are cleaned in the chocolate factory. Then the beans are roasted. Roasting the beans cooks them. It is easier to take shells off beans that have been roasted.

13

Machines mash the beans.

The shells are taken off the beans. Then the beans are mashed. Mashing the beans turns them into a very soft paste called **cocoa butter.**

The chocolate is mixed.

Milk and sugar are added to the cocoa butter to make chocolate. The chocolate is heated and mixed for several days. Mixing makes the chocolate smooth and creamy.

The chocolate is poured.

The chocolate is poured into **molds.** Molds are containers that are used to shape things. The chocolate is cooled in the molds. It becomes hard.

The chocolate is wrapped.

The chocolate is taken out of the molds. Machines wrap the chocolate. Trucks take the wrapped chocolate to stores to be sold.

I eat my favorite treat!

How many chocolate treats can you name? All of the chocolate in them started as cocoa beans!

Glossary

cocoa beans (KOH-koh BEENZ): seeds of a cocoa tree

cocoa butter (KOH-koh BUH-tuhr): a soft paste made from mashed cocoa beans

factory (FAK-tuh-ree): a building where things are made

molds (MOHLDZ): containers used to shape chocolate

pods (PAHDZ): fruits of a cocoa tree

24

Index